# SEAWEEDS OF AUSTRALIA

# SEAWEEDS OF AUSTRALIA

Photographs by Bruce Fuhrer

Text edited by I.G. Christianson,
M.N. Clayton, B.M. Allender

REED

*First published 1981*

A. H. & A. W. REED PTY LTD
2 Aquatic Drive, Frenchs Forest, Sydney, NSW 2086
68–74 Kingsford-Smith Street, Wellington 3
*also at*
Auckland and Christchurch



National Library of Australia
Cataloguing-in-Publication Data:

Seaweeds of Australia

Includes index.
ISBN 0 589 50293 X.

1. Marine flora—Australia. I. Fuhrer, B.A. (Bruce Alexander).

581.994

*Set by ProComp Productions Pty Ltd, Adelaide*
*Printed and bound by Kyodo Shing Loong, Singapore*

# CONTENTS

# FOREWORD

The beaches of Australia, justly famous, are enjoyed by her inhabitants for their lack of intellectual delight and stimulation: an elemental space of sunlight, clear water and sterile sand, soothing to minds oppressed by habits of busy-ness. Yet the more restless, at length satiated with sun and idleness, will wander to the adjacent headland in search of variety, fish to be inveigled, anemones to be deceived, crabs to be surprised, and the customary trophies of the shore, some durable, portable fragments of the sea's beauty in shell or stone. Of all the happy impressions of summers by the sea how many concern the plants of the sea? Indeed, if they are noticed at all it will probably be as blemishes of the seascape, decaying and slightly offensive heaps of jetsam, entanglements for lost fish-hooks. No kind of plant, if I may paraphrase a great botanist, is more beautiful than another; though some kinds need more training than others for their appreciation. This book is an attempt by a group of Australian botanists to help you see there is as much beauty to be found among the plants of that headland as in the forests or fields or any garden.

The beginning of wisdom is to be able to make distinctions. Those who are starting from the position of lumping all submarine vegetable matter together under the name of 'seaweed' are in good company. The great Linnaeus went little further. He distinguished initially only three groups: *Fucus* for all the massive ones, *Ulva* for all the flat ones, and *Conferva* for all the wispy ones. He did not recognise the division into three major colour classes, brown, green and red, which is the basis of all modern classification. To make finer distinctions between Australian algae has been, until the production of this book, a heavy task. ('Alga' is the singular with a hard g; 'algae' is the plural with the g soft as in 'genius'; it is the general name for water plants reproducing by simple—and obscure—structures, not flowers.) Most of the popular books about marine life are concerned solely with animals, or give only passing reference to a few common and conspicuous sea plants. Those books describing Australian seaweeds in detail are inaccessible to the public today. Harvey's *Phycologia Australica*, five volumes of descriptions and exquisite hand-coloured engravings published between 1858 and 1862, and works by other nineteenth-century botanists are collectors' items. *The Seaweeds of South Australia*, Part 1 (1936), by Lucas, and Part 2 (1947), by Lucas and Perrin, are long since out of print and written for experts. Names in all these are likely to be obsolete. The results of the considerable volume of work done on our marine plants in the last thirty years are scattered in specialist journals. The authors of this book have aimed to help the seaside visitor recognise and name seaweeds on the shore without the use of a microscope and other elaborate methods of examination which phycologists use in their laboratories. ('Phycologist' is the name for an expert on seaweeds—the same word as Harvey used in his title already mentioned, and derived from φυκοσ and the Latin *Fucus* of Linnaeus. It is a word to be used with caution—it is easily confounded by the careless with several other sciences whose names contain various assortments of p, s, h and y.)

The book is a joint project by members of the newly-established Australasian Society for Phycology and Aquatic Botany. It is some indication of the difficulties already referred to that the co-operative efforts of many people with different specialised skills have been necessary to produce this apparently simple book. Exploration and collection of plants around the coastline of the continent in different seasons has occupied students and professionals wading and diving over a number of years. The finding of valid names for the specimens demands both researches in the libraries and collections of Europe where most of the early work was done, and laboratory studies (often of extended duration) growing the plants and inducing them to form the reproductive structures on which their classification is based, studying the structures by light and electron microscopy. The presentation of the findings as coloured photographs (the modern equivalent of Harvey's engravings) which may be matched with the living plants and which obviate the need for learning the special language of technical description, has called for wholly new photographic techniques.

In early stages of the project several students at Monash University contributed notably to this development, especially Jenny Powell and Jan Aldenhoven. Bruce Fuhrer, whose colour photographs of Victorian land plants are renowned, has for some years extended his enthusiasm to the more challenging task of producing likenesses of seaweeds. This collection of pictures bears witness to his success. The problems of displaying, lighting and photographing seaweeds in water have not been better solved by anyone.

The task of identification and of the writing of the text and captions accompanying the photographs has been shared among a number of professional phycologists in Australian universities, with specific contributions from Drs Gerry Kraft and Sophie Ducker (Melbourne), Dr Bill Woelkerling (La Trobe), Mrs Julie Phillips (Monash) and Dr Ian Price (James Cook). Other members of the Australasian Society for Phycology and Aquatic Botany have given helpful informal advice, especially the President, Professor Bryan Womersley of Adelaide. The main part of the text and the coordination of the project are the work of members of the Botany Department of Monash University, Dr Margaret Clayton, Mrs Iona Christianson and Dr Bruce Allender, with whom lies editorial responsibility.

The present selection of 186 plates includes seaweeds found on our southern shores from Western Australia to New South Wales, including Tasmania, and also a few tropical species from Queensland. It represents only a small part of the flora to be found on these shores, and is intended as a forerunner to other volumes.

*M. J. Canny*
*Professor of Botany*
*Monash University*

# INTRODUCTION

The term 'seaweed' refers to the majority of the macroscopic plants which inhabit the intertidal region of the seashore and the permanently submersed shallow sunlit regions of the continental shelf. Seaweeds are algae: structurally simple plants without roots, stems, or leaves, and having primitive methods of reproduction. As well as the algae, there are several species of flowering plants, known as sea-grasses, found in coastal waters.

Seaweeds and sea-grasses are the major primary producers of coastal regions. Some coastal areas (for example, kelp beds and estuaries) surpass the richest agricultural land in the amount of plant material produced annually.

The macroscopic algae, along with the microscopic algae including phytoplankton, are the basis of many food webs. They are eaten by vertebrate and invertebrate animals, including some fish, sea urchins, limpets, and periwinkles. In contrast, the sea-grasses contribute to the productivity of coastal regions in a different way: few animals eat them, and they are left to decompose as a result of the activity of fungi and bacteria. These micro-organisms are themselves the food for many animals. Living seaweeds also 'leak' out various organic compounds into the surrounding seawater. Thus, in these various ways, its plants are vital to all life in the sea.

Just as terrestrial plants are restricted to certain positions or climates, so too are sea plants. The recognition of major types of seaweed habitats will thus be of assistance in identifying the plants. Especially on rocky shores, most species of seaweeds and marine animals have relatively restricted vertical distributions, occurring in horizontal bands or zones which are related to tidal fluctuations. Three zones have been recognised on coasts throughout the world, and although their vertical extent may vary they are always characterised by particular kinds of organisms (see Figure).

The majority of seaweeds occur around low tide level or subtidally. Pools on rock platforms may support the growth of plants which are normally found in much deeper water, with shade-loving species hidden under the pool ledges. Characteristically, seaweeds grow firmly attached to solid surfaces, but very many species may grow as epiphytes, that is, attached to other plants. Some are invariably epiphytic, such as *Notheia anomala* on *Hormosira banksii* (Plate 94). Along sheltered sandy coastlines the number of species is much lower, as is the range of habitats. It is here the sea-grasses reach their greatest abundance, and many species of algae grow on the sea-grasses as epiphytes.

Probably the first step in learning to identify seaweeds is to recognise the major groupings of species. The larger marine algae fall into three groups, known as divisions. The division to which an alga belongs can be determined easily on most occasions by its colour: red (Rhodophyta), brown (Phaeophyta), or green (Chlorophyta). A fourth division, numerically less important on the seashore, and also less conspicuous, consists of the blue-green algae (Cyanophyta). Frequently, blue-green algae appear to be very dark, almost black. Sometimes the colour of a specimen is not an absolutely reliable guide to its division. For example, Rhodophyta which grow in the intertidal zone do not look red, rather, they are usually brownish or even yellow.

Within these divisions there are further groupings of species which have an overall similarity based on their morphology. Botanical classification is hierarchical and there are several ranks of groups below that of division, including: CLASS, ORDER, FAMILY, GENUS and SPECIES. A species has two names—the first name is the genus and the second is the species. *Hormosira banksii* is the scientific name for the familiar 'Neptune's Necklace' of temperate Australian shores. It belongs to the DIVISION Phaeophyta, the ORDER Fucales (which includes most of the large brown seaweeds), and the FAMILY Hormosiraceae. It is a distinctive alga and *Hormosira* is the only genus in this family, with *H. banksii* the only species in the genus. The great majority of seaweeds are known only by their scientific names; some which have a history of usefulness in the culture of people living near a shore may have a common name, for example, 'Kelp' described several species of large brown algae which in Europe were burnt for the useful substances in their ash.

The scientific study of all groups of algae involves detailed studies of form, structure and development, invariably with the aid of a microscope. However, many species may be recognised by their macroscopic characteristics, and certainly the genus is often obvious, although to discriminate the species might entail a more difficult investigation.

As in all plant identification, with practice one develops a 'feel' for the plants and correctly grouping the specimens becomes easier.

**Above:** Plate 1
Extensive, near-horizontal, rocky platforms of dune limestone are common along south-eastern Australian coasts. The dominant plant of the lower intertidal zone is the brown alga *Hormosira banksii.*

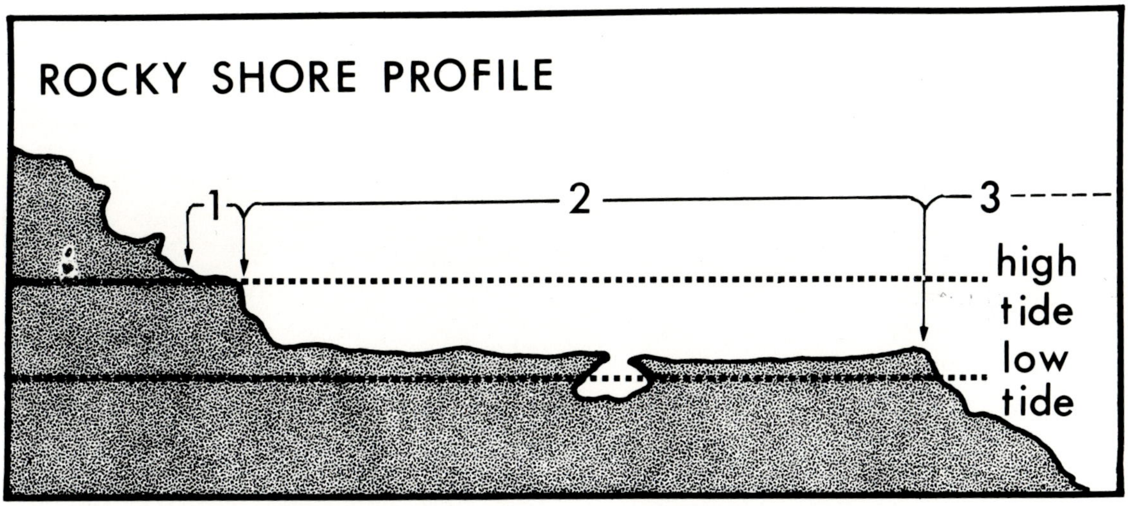


| Three major zones | Characteristic organisms | Approximate tide levels |
|---|---|---|
| 1 Littoral fringe | Periwinkles; black lichens and blue-green algae | Splash (above high tide) |
| 2 Eulittoral zone | Barnacles and limpets | Intertidal (between low and high tide) |
| 3 Sublittoral zone | Kelps (corals) | Subtidal |

(In the notes accompanying the photographs in this book, the terms 'Intertidal' and 'Subtidal' are used to describe seaweed habitats.)

**Top:** Plate 2

'Tide pools' or 'rock pools' provide a refuge for seaweeds that otherwise live in deeper waters. The rock surface around this pool supports only a growth of barnacles, but within the pool a varied flora is found. There are plants of the green *Ulva* sp. and *Codium fragile*, the brown *Cystophora moniliformis* in the centre, and an abundant shrubby growth of pink calcareous algae, in particular *Corallina* sp. Other coralline algae form crusts over the submersed rocks.

**Bottom:** Plate 3

This pool is at a lower level of the intertidal zone. Encrusting coralline algae again cover the rocky floor, but the larger algae are the brown seaweeds *Ecklonia radiata* and species of *Sargassum* and *Cystophora*.

# COLLECTION AND PRESERVATION OF SEAWEEDS

To study and collect seaweeds one must go to the shore when the tide is falling, preferably at a period of low-water spring tides, when the greatest number and variety of plants are exposed. During the Australian summer many intertidal seaweeds may dry up and disappear from the shore; the best growth is found in spring and early summer. Deep rock pools will usually have an abundant flora all the year, as will the subtidal zone which is best explored by diving. Seaweeds cast ashore on the beach dry out or decompose quickly, although interesting specimens from deep water may sometimes be found in storm drift.

Equipment needed for collecting trips includes a bucket, plastic bags or jars, a spoon or knife for scraping complete plants from the rocks, card labels and a pencil, and canvas shoes or wading boots that do not slip on wet rock.

Collection of specimens should be careful and selective, not wasteful or destructive, as once removed from the sea the plants soon decompose. Healthy and, where possible, entire plants should be chosen, and only those which can reasonably be preserved. Specimens need to be kept moist and cool, and processed as soon as possible after collection. Most seaweeds can be preserved satisfactorily as dried herbarium specimens, and some of the more delicate forms reveal an exquisite beauty of design when prepared carefully. These can also be used decoratively on greeting cards, or as framed wall hangings. Materials to be assembled for this process are sheets of firm white paper or card, pieces of cheese cloth or similar thin material cut to match the size of the paper, two or more shallow large dishes or trays, a large stack of newspaper, a plant press or a heavy flat object, forceps, scissors and fine brushes.

An individual specimen is spread in a shallow dish of seawater, cleaned of sand, small animals and epiphytic algae, and if very bushy it can be carefully pruned of some fronds so that the form is seen more clearly. The specimen is then floated and arranged over a sheet of the mounting paper in a tray of fresh water, and the paper is slowly tilted and removed from the tray, allowing the water to drain off and the seaweed to remain spread over the paper. It is then covered with a piece of cheese cloth, and pressed between several sheets of newspaper. A stack of such specimens is built up and put in a press. The newspaper must be changed frequently during the first day or two, and then daily until the specimens are quite dry. Moulds develop if drying is slow. Steady firm pressure is essential to obtain flat unwarped sheets. Delicate soft algae adhere very firmly to the mounting paper and less closely to the cheese cloth cover which can be peeled off when drying is completed. Coarse, thick or wiry algae do not adhere satisfactorily but when dry can be attached to the paper with glue or strips of gummed paper.

Date and site of collection, collector's name, and other relevant collection data should be recorded on each herbarium specimen. This information or a code number should be pencilled on the dry paper before the plant is mounted, and a permanent label affixed later.

Dried specimens, if stored flat and away from light, will keep indefinitely and not lose their colour. Small specimens or parts of plants to be examined microscopically can be preserved in labelled vials containing 4 per cent formalin seawater, kept in the dark.

# BLUE-GREEN ALGAE
## *(Cyanophyta)*

The marine representatives of this division are often overlooked as 'seaweeds' because they are usually small and inconspicuous, although they may cover extensive areas in habitats which the main seaweed groups cannot tolerate.

A slimy skin or crust exposed at low tide on calm mud flats or on rocks of tropical reefs usually consists of various blue-green algae. On rock faces of temperate coasts exposed to strong wave action blue-green algae form black patches or bands in the splash zone.

Colonies of blue-greens are also found in the temperate intertidal zone among limpets and barnacles. *Rivularia firma* forms firm gelatinous colonies which are slippery to touch; some other species of *Rivularia* are hollow and jelly-like to touch. *Symploca* forms low irregular patches which feel like soft hair. Both illustrations are from intertidal reef platforms in southern Australia.

**Above:** Plate 4 (x 1½) *Rivularia firma*

**Left:** Plate 5 (x ½) *Symploca* sp.

# RED ALGAE
## *(Rhodophyta)*

The red algae are the most numerous of the seaweed groups and have a stunning range of form, texture and colour. A few occur intertidally but most species, and certainly the more delicate and intricate forms, are subtidal, many of them in deep water and found only by diving or when cast ashore as drift. Some are found in tidal pools and under shaded ledges. The colour of subtidal plants is clearly red; intertidal species appear a yellow or brownish-red.

Southern Australian coasts are particularly rich in species and a high proportion of these fascinating and beautiful plants are endemic to this area; on tropical coasts red algae are less numerous.

The identification of red algae can be very frustrating as the features used to distinguish the species are mainly microscopic. As yet, many Australian species are poorly understood and some are unnamed.

Arrangement of the pictures in this book is according to orders (shown in parentheses), in the divisions of brown and green algae. For the red algae, however, the arrangement chosen is according to general appearance of the plants, as specimens which look very similar may belong to different orders. Conversely, some plants which are superficially quite dissimilar may be closely related. The name of the family as well as the order to which a red alga belongs is given where the family is large or fairly distinctive.

**Below**: Plate 6
The beach at Port MacDonnell (South Australia) is seen piled with drift seaweed after a storm. Many of the red algae photographed for this book were collected there. These deep-water species are bright pink to dark red or purple when growing, but change colour and bleach rapidly on the shore.

**Top:** Plate 7 (x 4)

*Amphiroa* sp.

(Corallinaceae, Cryptonemiales)

Australian coasts are rich in coralline algae, so called because they resemble animal corals both in appearance and in the fact that they are more or less impregnated with calcium carbonate. There are two kinds: those which form a crust, and those which have 'articulated' branches composed of calcified segments and uncalcified joints.

**Bottom:** Plate 8 (x 2)

*Amphiroa* sp.

(Corallinaceae, Cryptonemiales)

This genus of articulated coralline algae has flattened branches which fork at elaborate joints. The tips of the branches are the most actively growing part of the plant and appear white: further down, the branches become the typical pink colour of coralline algae.

**Top:** Plate 9 (x 4)

*Corallina officinalis*

(Corallinaceae, Cryptonemiales)

Species of the feather-like *Corallina* occur commonly in the tide pools and on lower intertidal rock surfaces. The rounded reproductive structures can be seen at the tips of some branches. This is a cosmopolitan species: the name *officinalis* means that the plants were used by herbalists for medicinal purposes.

**Bottom:** Plate 10 (x 4)

*Cheilosporum sagittatum*

(Corallinaceae, Cryptonemiales)

In this common coralline of south-eastern Australia all the branches are flattened and each segment is shaped like an arrow-head. Young and very old plants may not show quite such a regular pattern.

**Top:** Plate 11 (x 2)

*Metagoniolithon stelliferum*
(Corallinaceae, Cryptonemiales)
This profusely branched and highly coloured coralline alga grows as an epiphyte on the marine flowering plant *Amphibolus antarctica.* The stems and leaves of the sea grass are sometimes so overgrown by seaweed that the branches become too heavy and are torn off the main plant.

**Bottom:** Plate 12 (x 1)

*Liagora harveyana* (Nemalionales)
Although calcified, this species is quite distinct from the Corallinaceae. Its forked branches are cylindrical, not jointed. Plants can occur commonly in rock pools in areas of moderate or rough wave action along southern coasts.

**Top**: Plate 13 (x 2)

Encrusting Corallinaceae
(Cryptonemiales)

Pink, stony, calcified crusts of many coralline algae form prostrate expanses over rocks, or on other plants. While extremely common along all Australian coasts, most kinds cannot be identified even to genus level without knowing details of their internal structure.

**Bottom**: Plate 14 (x 2)

The reproductive structures are often inconspicuous or sunk within the crust. The white dots on this pink crust are the openings from these structures.

**Top:** Plate 15 (x 1)

Most encrusting corallines grow in shaded situations and often form nearly continuous carpets over the walls and floor of tide pools.

**Bottom:** Plate 16 (x 2)

Species which grow on rocks often become massive and can be over 1 cm thick. In the tropics, encrusting coralline algae are important in the formation of coral reefs.

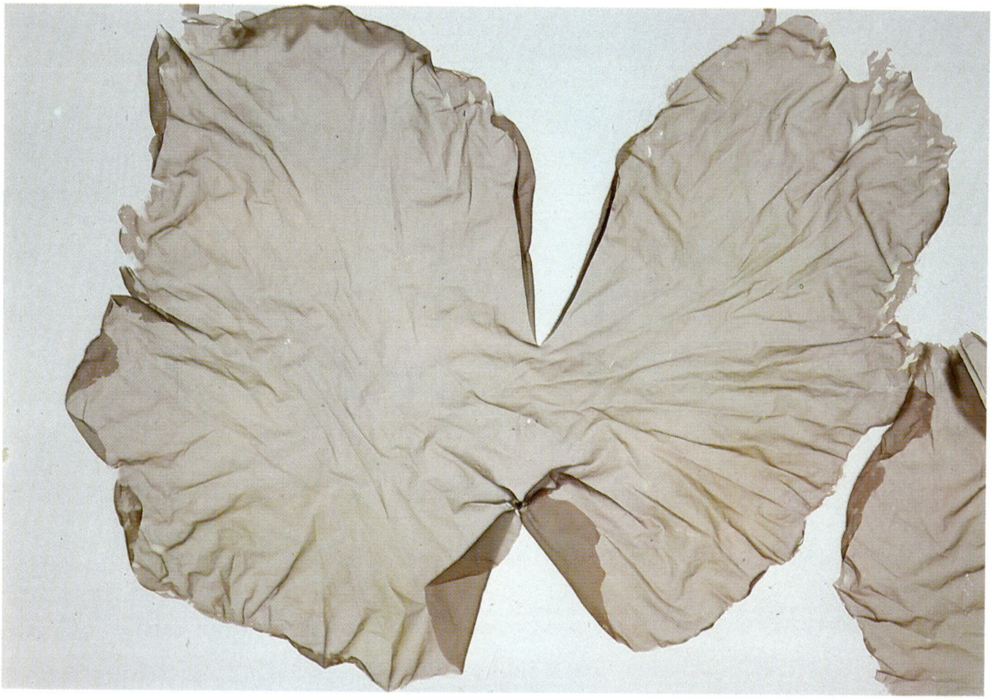

**Top:** Plate 17 (x 1)

*Porphyra lucasii* (Bangiales)
Species of *Porphyra* occur commonly in winter and spring on intertidal rocks of temperate Australian coasts. The thin membranous sheets vary in shape from ribbon-like to broadly oval, and in colour from a light bronze to purple or almost black. *Porphyra* is used as a food, particularly in Japan where it is known as *nori*. In the British Isles it is called laver.

**Bottom:** Plate 18 (x 1)

*Sonderophycus australis* (Cryptonemiales)
The fan-shaped leathery crusts are up to 15 cm across and have a woolly-looking under-surface which is attached to rock.

**Top:** Plate 19 (x 1/3)

*Gloiophyllis barkeriae* (Gigartinales)
Many genera of red algae have blade-like to wide, leaf-shaped fronds, and in Australia most plants of this form grow in deep water. This species, apparently rather rare, is known only from south-eastern Australia.

**Bottom:** Plate 20 (x 1/2)

*Schizymenia* sp. (Gigartinales)
Although similar in external appearance, these genera differ from one another in internal structure and reproductive details. *Schizymenia* has a relatively leathery consistency.

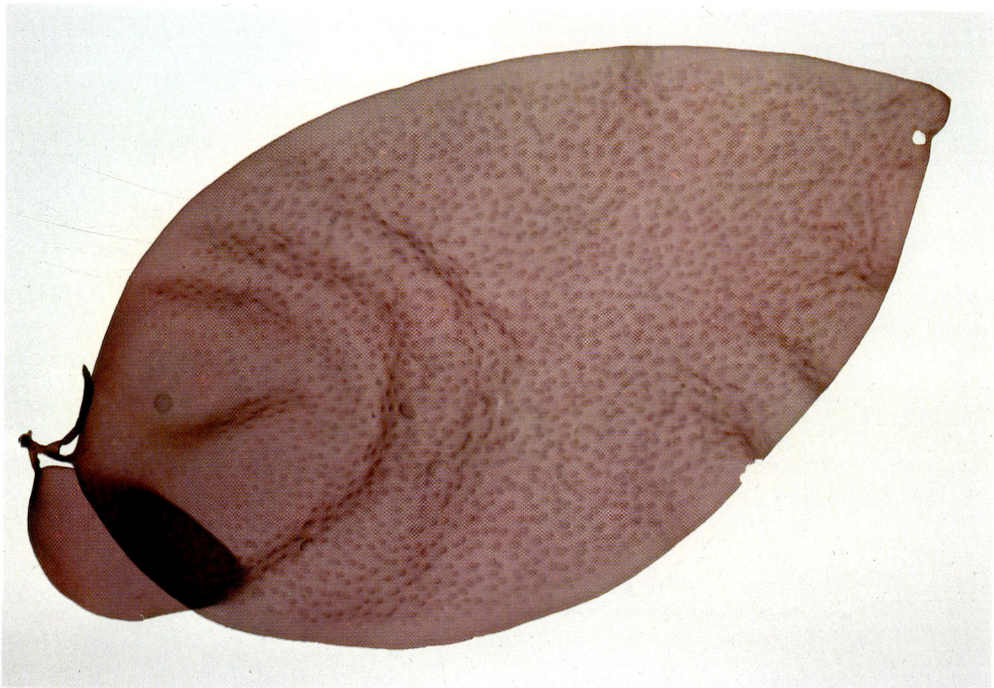

**Top:** Plate 21 (x ½)

*Rhodoglossum proliferum* (Gigartinales)

Species of this genus are also found in other parts of the world. *Rhodoglossum* is rather delicate and papery.

**Bottom:** Plate 22 (x ½)

*Gigartina radula* (Gigartinales)

The leather fronds of *Gigartina* vary tremendously in form. Some species, like *G. radula*, have large unbranched blades. This specimen bears conspicuous reproductive structures.

**Top:** Plate 23 (x 1)

*Gigartina wehliae* (Gigartinales)
The genus *Gigartina* is widely distributed in temperate waters, with several species in Australia. *G. wehliae* appears to be a rare plant, known only from the Port MacDonnell area in South Australia.

**Bottom:** Plate 24 (x ⅔)
*Gigartina wehliae*

**Top:** Plate 25 (x 1)

*Gigartina muelleriana* (Gigartinales)
Although the tough fronds of *Gigartina* are always flattened, the degree of branching varies considerably between the different species.

**Bottom:** Plate 26 (x ⅓)

*Gigartina crassicaulis* (Gigartinales)
All species of *Gigartina* have similar reproductive structures and thus are placed in the same genus.

**Top:** Plate 27 (x ½)

*Rhodymenia australis* (Rhodymeniales) Flattened fronds with fairly regularly forked branching are found in a number of red seaweeds.

**Bottom:** Plate 28 (x 1)

*Curdiea laciniata* (Cryptonemiales) The fronds are tough and leathery; spherical reproductive structures are borne on the margins. This is a fairly common plant in drift on southern coasts.

**Top:** Plate 29 (x 1)

*Hymenena curdieana*
(Delesseriaceae, Ceramiales)
Most genera of the Delesseriaceae have a thin leafy form and delicate appearance. In some, such as *H. curdieana*, the blades show a series of very fine 'veins'.

**Bottom:** Plate 30 (x 2)

*Hymenena curdieana*
Reproductive structures appear as dark spots on the 'veins' of the leafy fronds. A rose-pink colour is typical for the Delesseriaceae.

**Top**: Plate 31 (x 2)

*Heterodoxia denticulata*
(Delesseriaceae, Ceramiales)
The 'veins' are conspicuous and form ridge-like thickenings along the fronds. Members of the family Delesseriaceae usually grow in deep water and often deteriorate quickly when washed ashore.

**Bottom**: Plate 32 (x ⅔)

*Hemineura frondosa*
(Delesseriaceae, Ceramiales)
In this species 'veins' are scarcely apparent. All three species pictured here occur in temperate waters.

**Top:** Plate 33 (x 1)

*Craspedocarpus ramentaceus*

(Gigartinales)

This species of south-eastern Australia grows up to 20 cm in length. Margins of the larger branches of the frond often have small lateral proliferations 1–2 mm wide.

**Bottom:** Plate 34 (x 1½)

*Dictymenia tridens*

(Rhodomelaceae, Ceramiales)

There are several species of *Dictymenia* in temperate waters of Australia. They have a network of vein-like markings on the narrow blades and a regular branching pattern. The three-pronged tooth-like margins are characteristic of *D. tridens.*

**Top:** Plate 35 (x 1)

*Jeannerettia pedicellata*
(Rhodomelaceae, Ceramiales)
Plants reach 12 cm in length, and are distinguished from the following species by the absence of midribs on the blades, although radiating veinlets are seen on the margins of the fronds.

**Bottom:** Plate 36 (x ⅔)

*Jeannerettia lobata*
(Rhodomelaceae, Ceramiales)
Plants of this species grow to lengths of 50 cm and are quite common in the drift on southern shores. The membranous fronds have a distinct midrib.

**Top:** Plate 37 (x 1½)

*Lenormandia marginata*
(Rhodomelaceae, Ceramiales)
The genus can be recognised by the leafy fronds which branch from the margins, or a midrib, or both, and they have fine lines on them, as seen in this plant of *L. marginata*.

**Bottom:** Plate 38 (x 1½)

*Lenormandia marginata*
Species of *Lenormandia* appear to be endemic to temperate Australian waters where they grow subtidally.

**Top:** Plate 39 (x 1)

*Lenormandia spectabilis*
(Rhodomelaceae, Ceramiales)
Colour can vary considerably in the species of *Lenormandia.* This can be due to differences in age or in available light, or to the onset of decomposition of drift specimens, as evident in this specimen of the broad-leafed *L. spectabilis.*

**Bottom:** Plate 40 (x ½)

*Lenormandia muelleri*
(Rhodomelaceae, Ceramiales)
Plants of this species can grow to 60 cm in length and are firmer in texture than the others illustrated.

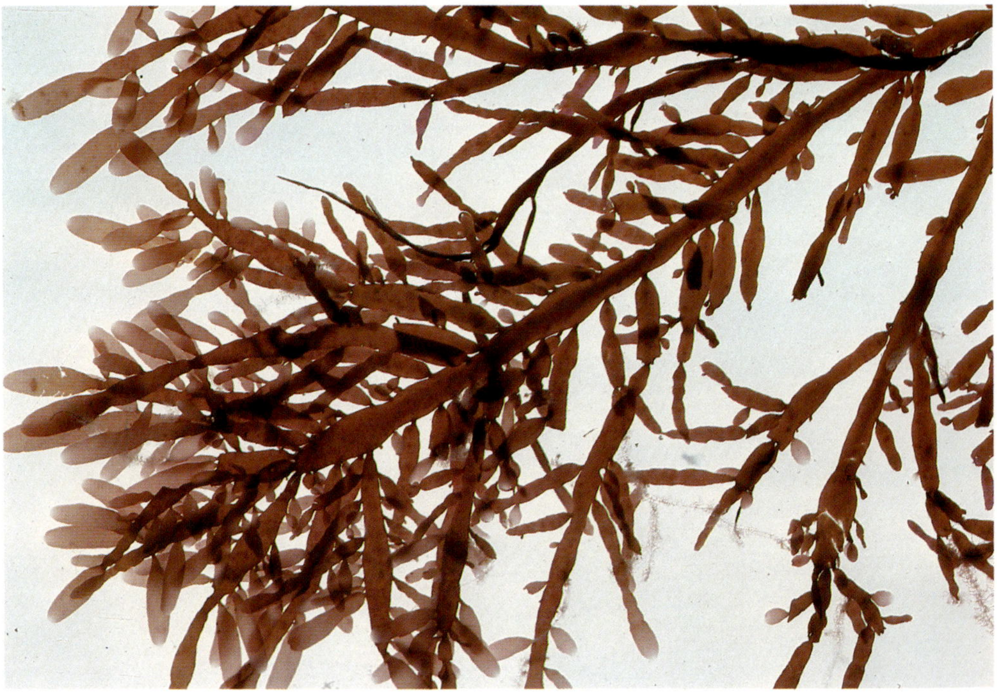

**Top:** Plate 41 (x 1½)

*Pterocladia lucida* (Gelidiales)
This much-branched alga can grow to 60 cm in length. Plants of the genus *Pterocladia* are a source of the gel-forming substance agar. This species occurs in southern Australian waters.

**Bottom:** Plate 42 (x 1)

*Nizymenia australis* (Gigartinales)
Plants of this common species are up to 20 cm in length. The outline is rather like *Lenormandia* but the substance is firmer and denser.

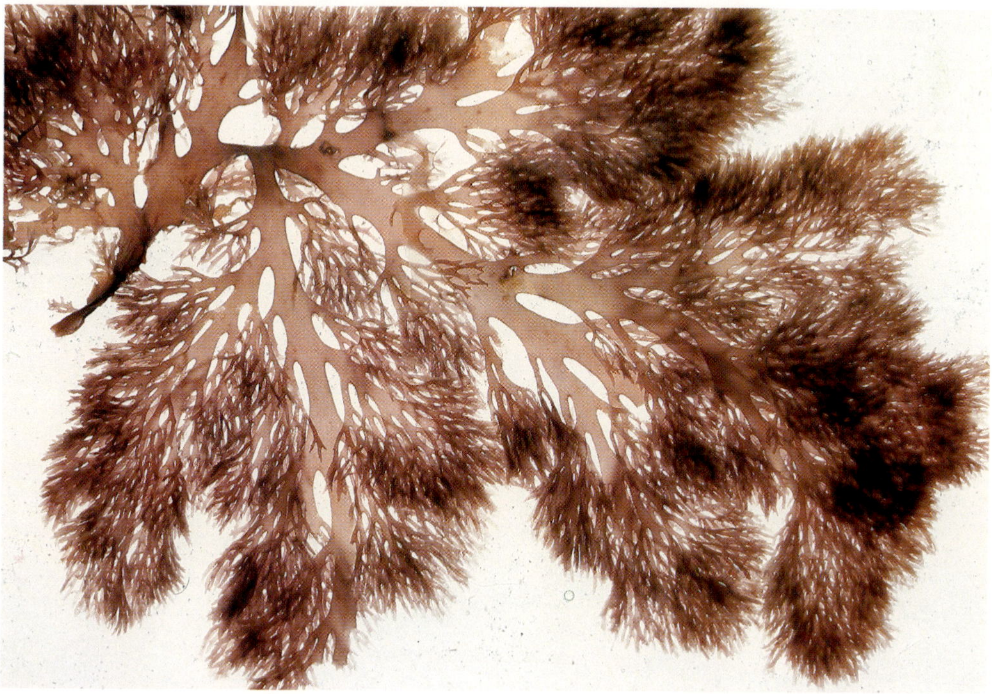

**Top:** Plate 43 (x 1)

*Callophyllis lambertii* (Cryptonemiales)
Specimens of this plant, which has an easily recognised branching pattern in one plane, are common in the drift on southern shores. It has a firm substance, is about 1 mm thick, and can grow to 40 cm in length.

**Bottom:** Plate 44 (x 1)

*Callophyllis rangiferinus* (Cryptonemiales)
The branching pattern here is basically the same, but the difference in width between the main branches and the end branches is very much greater than in *C. lambertii.* It is a smaller and more slender plant, usually less than 15 cm high.

**Top:** Plate 45 (x 5)

*Plocamium angustum* (Gigartinales)
The genus *Plocamium* is easily recognised by the distinctive branching pattern of its flat fronds. Branching is in one plane, with alternating series of two or more small branchlets from the edges of the main branches. The lower of these branchlets is itself usually unbranched, and is tooth-like or claw-shaped. In *P angustum* the branchlets alternate in pairs. The main branches are narrow, rarely more than 1 mm wide.

**Bottom:** Plate 46 (x 2)

*Plocamium dilatatum* (Gigartinales)
The branchlets alternate in pairs, and the outer edge of the wide claw-shaped lower branchlet is serrated. The main branches are relatively thick and commonly more than 2 mm wide.

**Top:** Plate 47 (x 5)

*Plocamium cartilagineum* (Gigartinales)
In this species the branchlets occur in alternating series of three (or four). *Plocamium* is a world-wide genus, with eight species recorded for southern Australia. Species of *Plocamium* often grow abundantly in the subtidal zone and are very common as drift on the shore.

**Bottom:** Plate 48 (x 1⅓)

*Plocamium mertensii* (Gigartinales)
This species also has broad main branches, up to 5 mm, but they are thin. The branchlets alternate in pairs and sometimes have a tufted appearance. This specimen has reproductive structures clustered in the forks of the branches.

**Top:** Plate 49 (x 3)

*Phacelocarpus alatus* (Gigartinales)
The long firm branches of this dark red alga are somewhat flattened and along their edges have rows of tooth-like projections. The teeth are shorter than the width of the branch. This species is found in temperate and tropical Australia.

**Bottom:** Plate 50 (x 3)

*Phacelocarpus labillardieri* (Gigartinales)
The branches of this species are rounded and bear rows of flattened tooth-like projections which are no longer than the width of the branch. Reproductive structures develop between these projections. *P. labillardieri* grows subtidally in temperate waters.

**Top:** Plate 51 (x 1)

*Delisea pulchra* (Nemalionales)
Plants of this species superficially resemble *Phacelocarpus*, but the internal structure is different. A second species of *Delisea*, appearing rather unlike *D. pulchra*, is shown in Plate 69.

**Bottom:** Plate 52 (x 3)

*Delisea pulchra*
The reproductive structures are borne near the tips of the branches and, if present, at once distinguish the species from *Phacelocarpus*. *D. pulchra* occurs in temperate and subtropical waters.

**Top**: Plate 53 (x 2)

*Gelidium glandulaefolium* (Gelidiales)
*Gelidium* is a cosmopolitan genus of many species, some of which are harvested for the production of the gel-forming agar. *G. glandulaefolium* is found in south-eastern Australia, where plants grow to over 25 cm in height. The branches commonly bear small glandular proliferations.

**Bottom**: Plate 54 (x 3)

*Gelidium australe* (Gelidiales)
Flat fronds arise from the base of the plant, and branch several times in a feather-like form. This species grows to about 8 cm high, and occurs in tropical and temperate regions of Australia.

**Top**: Plate 55 (x 1)

*Gelidium pusillum* (Gelidiales)
This species forms clumps or extensive turfs, with a crisp, rather brittle texture, on intertidal rock platforms in temperate Australia and elsewhere. Like the few other red algae found in the intertidal zone, the plants do not appear red, but yellow to reddish brown in colour.

**Bottom**: Plate 56 (x 5)

*Gelidium pusillum*
As in all species of *Gelidium*, the stiff fronds are somewhat flattened and bear branches from their margins in one plane. The tangled tufts of *G. pusillum* may be no more than 1 cm high.

**Top**: Plate 57 (x 1½)

*Pterocladia pinnata* (Gelidiales)
The form of this plant is similar to that of species in the related genus *Gelidium*. The texture is firm and coarse.

**Bottom**: Plate 58 (x 1)

*Grateloupia filicina* (Cryptonemiales)
The texture of this plant is firm but the surface is very slippery. *G. filicina* is a common species in sheltered bays of temperate and subtropical regions, where it grows to 25 cm in length.

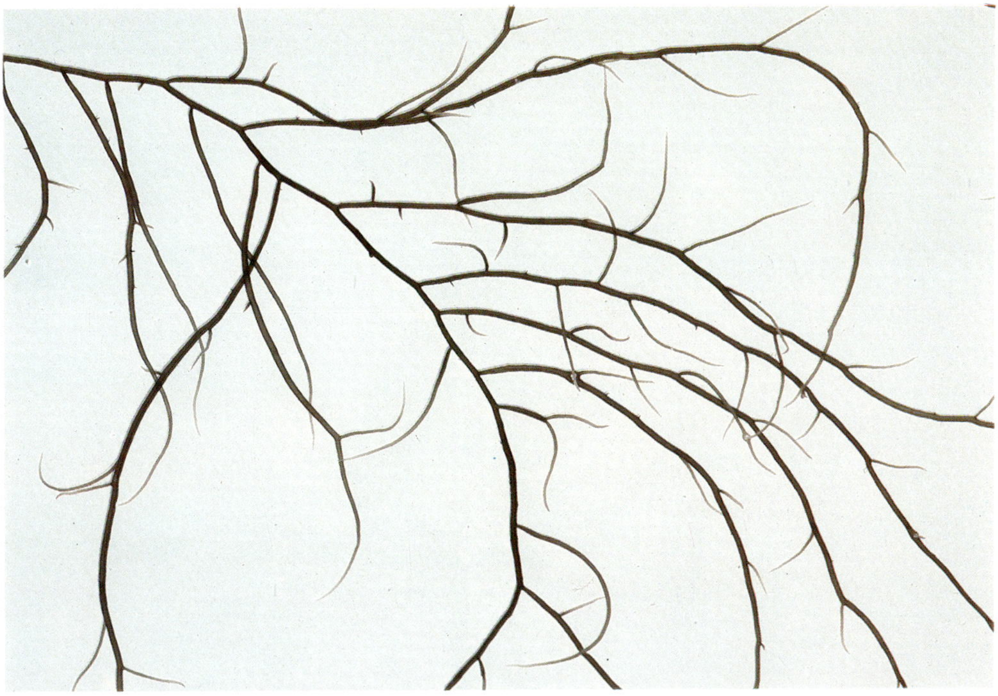

**Top:** Plate 59 (x 1)

*Gracilaria secundata* (Gigartinales)
This is one of the commonest species of *Gracilaria* in Australia, being found in rock pools, subtidal areas of open coasts, and estuaries. It has branching mostly from one side of the main branches. *Gracilaria* was harvested in New South Wales during the 1940s for the extraction of agar.

**Bottom:** Plate 60 (x 2)

*Hypnea episcopalis* (Gigartinales)
The hooked appendages formed at the end of some of the cylindrical branches are a useful though not always reliable guide in identifying this species. It is found in temperate and tropical Australia in subtidal habitats, and grows to 40 cm in length.

**Top:** Plate 61 (x 1)

*Nemastoma feredayae* (Gigartinales)
This deep-water alga of south-eastern Australia has an extremely gelatinous and slippery texture. This specimen is 15 cm high, and larger plants can be found.

**Bottom:** Plate 62 (x 1)

*Lomentaria australis* (Rhodymeniales)
Plants of this species grow to 10 cm or more in length in temperate subtidal environments. They have a soft gelatinous consistency and often deteriorate quickly after washing ashore in the drift.

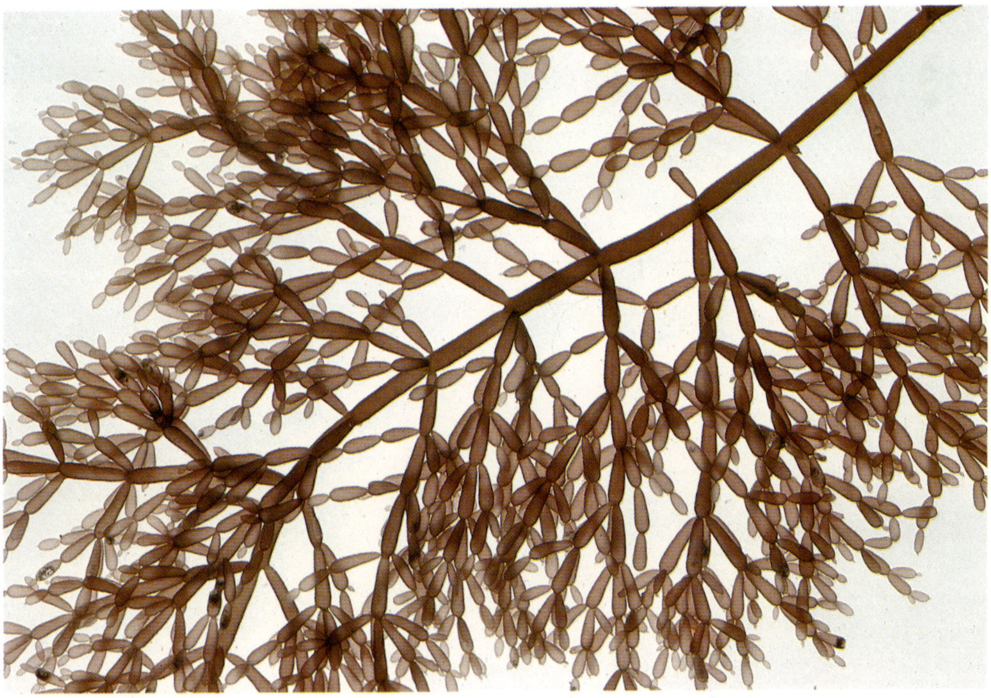

**Top:** Plate 63 (x ⅔)

*Solieria robusta* (Gigartinales)
This is another deep-water species of temperate Australia. The cylindrical or slightly compressed branches have a soft, fleshy consistency.

**Bottom:** Plate 64 (x 2)

*Erythroclonium muelleri* (Gigartinales)
Branches consisting of a series of cylindrical, more or less rounded or constricted segments are found in several genera of red algae. *E. muelleri* occurs in deep pools and subtidally. This plant is 10 cm high.

**Top:** Plate 65 (x 1½)

*Champia viridis* (Rhodymeniales)
The branches of *Champia* are partitioned by diaphragms into a series of distinct segments; usually this character readily distinguishes the genus. *C. viridis* is a species of temperate regions found in deep tide pools and subtidally. It grows to 15 cm or more in height.

**Bottom:** Plate 66 (x 2½)

*Botryocladia obovata* (Rhodymeniales)
This distinctive plant with clustered hollow grape-like branches grows in deep shaded tidal pools and subtidal environments. It is common in temperate Australia.

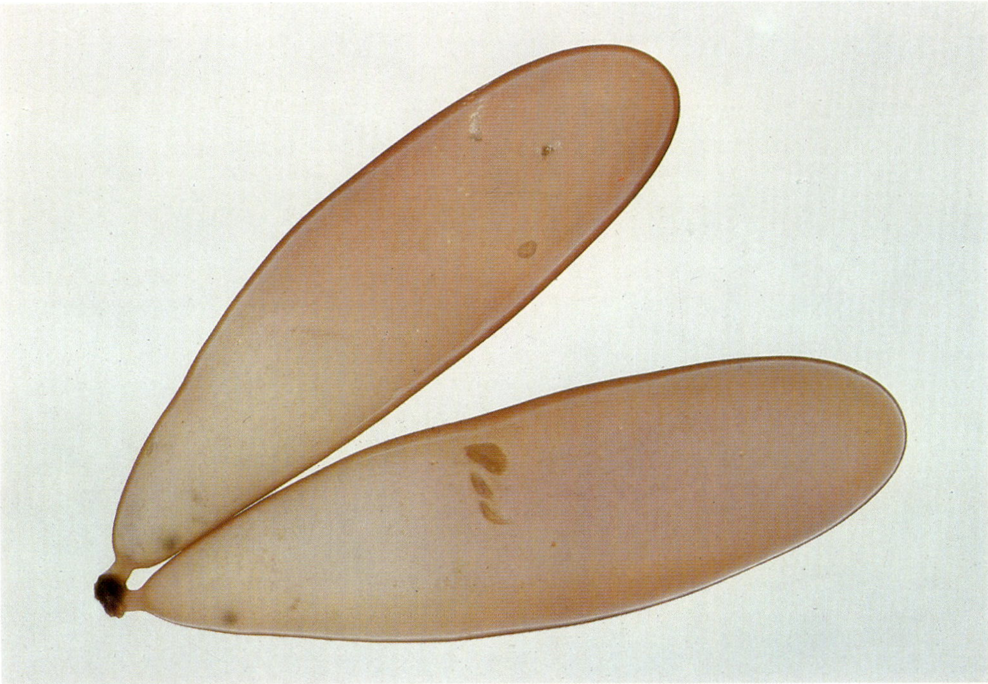

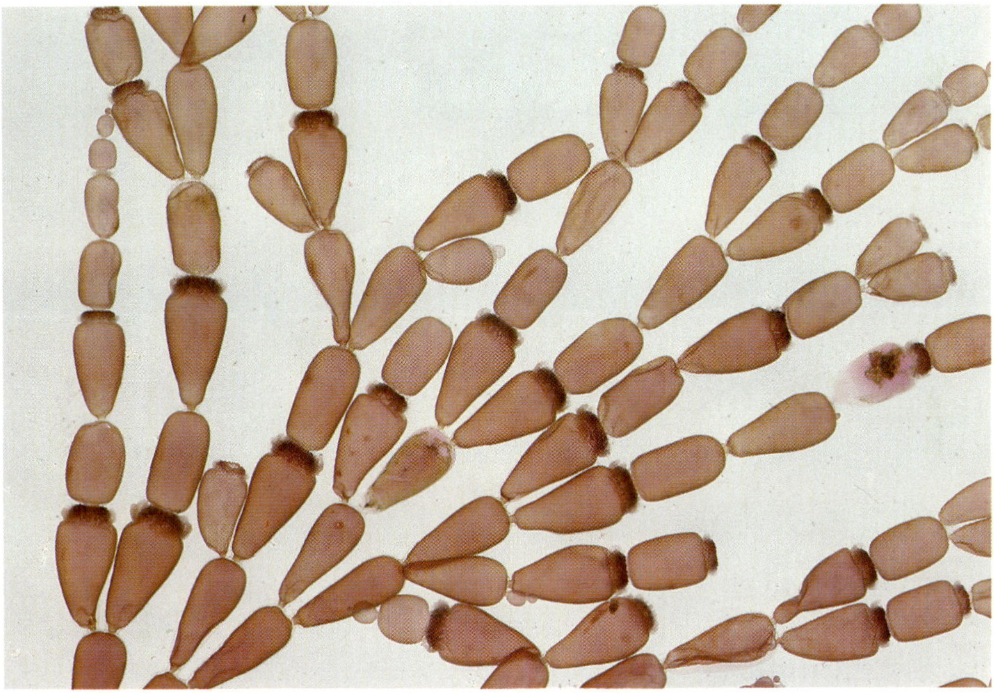

**Top:** Plate 67 (x 1)

*Botryocladia brownii* (Rhodymeniales)
Several hollow sausage-like fronds arise from the small holdfast, and can be up to 30 cm high. The interior is filled with a clear jelly. This is a species of subtidal environments along southern Australian shores.

**Bottom:** Plate 68 (x 3)

*Griffithsia ovalis*
(Ceramiaceae, Ceramiales)
The rounded segments of the preceding genera are complex and multicellular in structure. In *Griffithsia* these are large single cells, up to 2 or 3 mm wide, forming branched chains of pink glistening beads. This species occurs subtidally and grows as an epiphyte in tufts up to 3 cm long.

**Top:** Plate 69 (x 2)

*Delisea hypneoides* (Nemalionales)
The fronds of *D. hypneoides* are slender, much branched and of soft consistency, in contrast to the firmer, broader fronds of *D. pulchra* (Plate 51). The ends of the branches can be seen to be similar, with reproductive structures developing there. Plants grow subtidally in temperate waters, and reach 50 cm in length.

**Bottom:** Plate 70 (x 2)

*Heterosiphonia gunniana*
(Dasyaceae, Ceramiales)
This is a very delicate, soft alga of deep tide pools and subtidal habitats. The branches characteristically terminate in a fan-like spread of small branchlets. It is found along temperate Australian coasts.

**Top:** Plate 71 (x ⅓)

*Dasya extensa* (Dasyaceae, Ceramiales)
Plants of this species reach lengths of over 1 m, growing subtidally in temperate and subtropical waters. The long cylindrical fronds are very slimy and hairy.

**Bottom:** Plate 72 (x 2)

*Dasya extensa*
Reproductive structures appear as small dots at this magnification. Without microscopic examination, proper identification is difficult in *Dasya* and related genera which have the same general appearance.

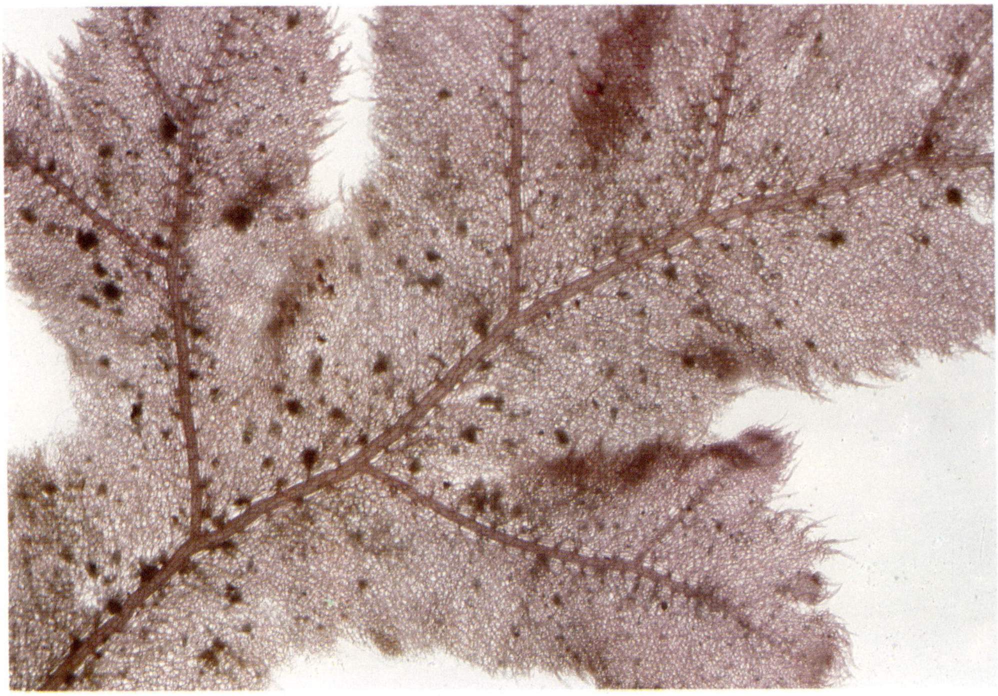

**Top:** Plate 73 (x 1)

*Thuretia quercifolia*
(Dasyaceae, Ceramiales)
The ultimate branchlets are interwoven to form a flattened network which vaguely resembles an oak leaf. This distinctive subtidal species grows to lengths of 25 cm, and is common in drift on temperate coasts.

**Bottom:** Plate 74 (x 2½)

*Thuretia quercifolia*
The network often becomes permeated with entrapped sand grains and other particles.

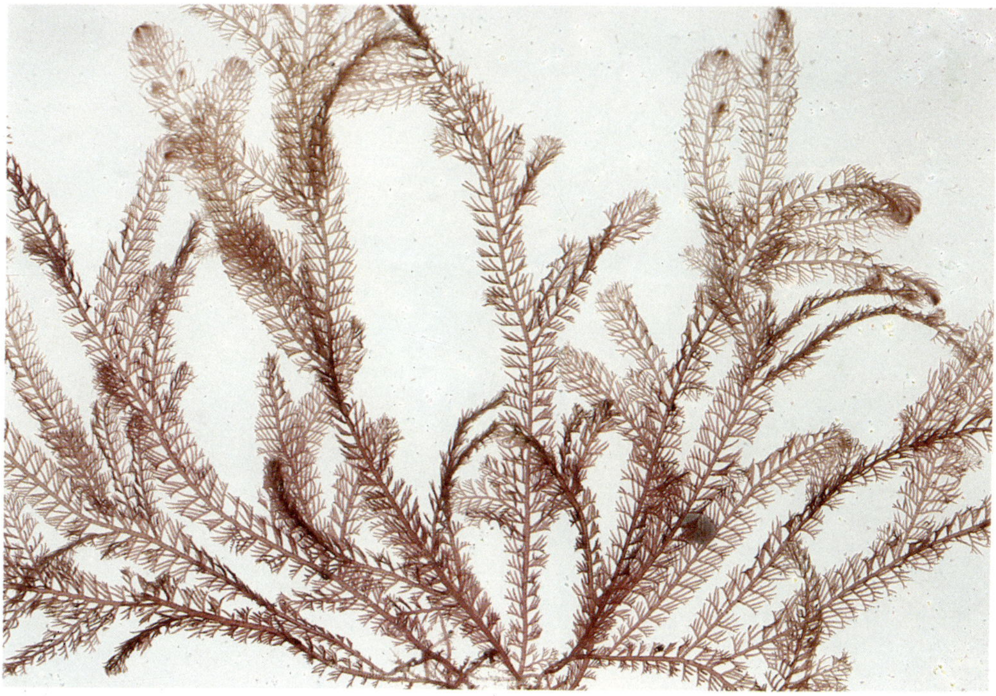

**Top:** Plate 75 (x 1)

*Haloplegma preissii*

(Ceramiaceae, Ceramiales)

The spongy consistency and felted appearance of this plant come from a densely interwoven network of filaments forming the flattened fronds which are of irregular shape, and up to 30 cm in length. Specimens are common in drift on temperate shores, but quickly lose their colour.

**Bottom:** Plate 76 (x 2)

*Dasyclonium incisum*

(Rhodomelaceae, Ceramiales)

This delicate species is usually epiphytic on other seaweeds. The tufts may be quite small or up to 10 cm in height. The distinctive branching pattern results in a comb-like feathery appearance. Plants are common in temperate waters.

**Top:** Plate 77 (x 3)

*Griffithsia* sp. on *Ptilota hannafordii*
(both Ceramiaceae, Ceramiales)
Species of *Griffithsia* have large cells often visible to the unaided eye (see also Plate 68); the frequent branching of the chains of cells gives the plants a delicate feathery appearance.

**Bottom:** Plate 78 (x 2)

*Ceramium* sp. (Ceramiaceae, Ceramiales)
The numerous species of this common cosmopolitan genus normally cannot be identified without the use of a microscope. The characteristic banded appearance of many Ceramiums is often just visible to the unaided eye. Plants may grow on rock or as epiphytes, in pools or subtidally.

**Top:** Plate 79 (x 1)

*Echinothamnion hystrix*
(Rhodomelaceae, Ceramiales)
There are many species and genera of the family Rhodomelaceae with a similar structure, found subtidally on open coasts and in sheltered bays. Their basically filamentous structure gives most of these plants a soft tufted or shrubby appearance.

**Bottom:** Plate 80 (x 2½)

*Echinothamnion hystrix*
Small tufts of branchlets distinguish this species.

**Top:** Plate 81 (x 1)

*Euptilota articulata*

(Ceramiaceae, Ceramiales)

This is a fairly common subtidal species of temperate coasts, growing to 30 cm or more in length.

**Bottom:** Plate 82 (x 2)

*Euptilota articulata*

Although the repeated feather-like branches are very fine, the substance of the plant is firm compared with most other species of Ceramiaceae illustrated here.

**Top:** Plate 83 (x 2)

*Ballia callitricha*

(Ceramiaceae, Ceramiales)

The close feather-like branching along the main branches and a rather stiff texture are distinguishing features of this species. It grows subtidally or sometimes in deep rock pools along temperate coasts, and is common in drift. Plants up to 20 cm or more in height can be found.

**Bottom:** Plate 84 (x 2)

*Wrangelia nobilis*

(Ceramiaceae, Ceramiales)

There are many species of *Wrangelia* and related genera on Australian coasts, not easily distinguished without microscopic examination. Numerous small branchlets give a woolly or hairy appearance to the rounded or compressed branches.

**Top**: Plate 85 (x 2½)

*Laurencia elata*

(Rhodomelaceae, Ceramiales)

Species of *Laurencia* are common on temperate and tropical coasts of Australia, in the upper subtidal zone and in rock pools. The genus is usually easy to recognise by an indentation or pit at the blunt tips of the branches. *L. elata*, common on temperate shores, can grow to 40 cm in length. Its firm, more or less compressed branches are 2–4 mm broad. Branching is mostly in one plane.

**Bottom**: Plate 86 (x 2½)

*Laurencia botryoides*

(Rhodomelaceae, Ceramiales)

Plants of this species are common on southern coasts, are usually a reddish brown colour, and are up to about 15 cm high. Branching is in one plane, but the side branches are not much flattened, and branchlets look clustered.

**Above:** Plate 87 (x 2)

*Laurencia filiformis*
(Rhodomelaceae, Ceramiales)
This species is also common in tide pools and subtidal habitats on temperate coasts. The fronds are cylindrical with more or less radial branching. Plants grow to about 15 cm, but in deep water may grow much taller.

## BROWN ALGAE
### *(Phaeophyta)*

The shades of colour seen in this group of seaweeds range from a pale olive- or yellow-brown to almost black.

The brown algae are the most conspicuous seaweeds in the lower intertidal and upper subtidal regions of temperate rocky shores. Many of the common species in Australia and New Zealand are found nowhere else in the world.

There is a wide range of size and form in this group, from thin crusts and small tufts to the massive kelps which rival forest trees in height.

Because of their large size, some of the brown seaweeds are illustrated in this book only by characteristic parts of the plants. Unless one is a diver many of the large brown seaweeds will only be found as drift on the beach.

**Below:** Plate 88

The flora of lower intertidal rock pools is often dominated by large brown seaweeds. This pool, fringed by *Hormosira banksii*, contains *Ecklonia radiata*, *Xiphophora chondrophylla*, and several species of *Cystophora* and *Sargassum*.

**Below:** Plate 89

A very low tide reveals some of the large brown seaweeds which are abundant in the upper subtidal zone of a rocky shore. *Durvillaea potatorum* and *Phyllospora comosa* are seen exposed, while fronds of *Macrocystis angustifolia* float in the water. The green *Codium fragile* is also conspicuous.

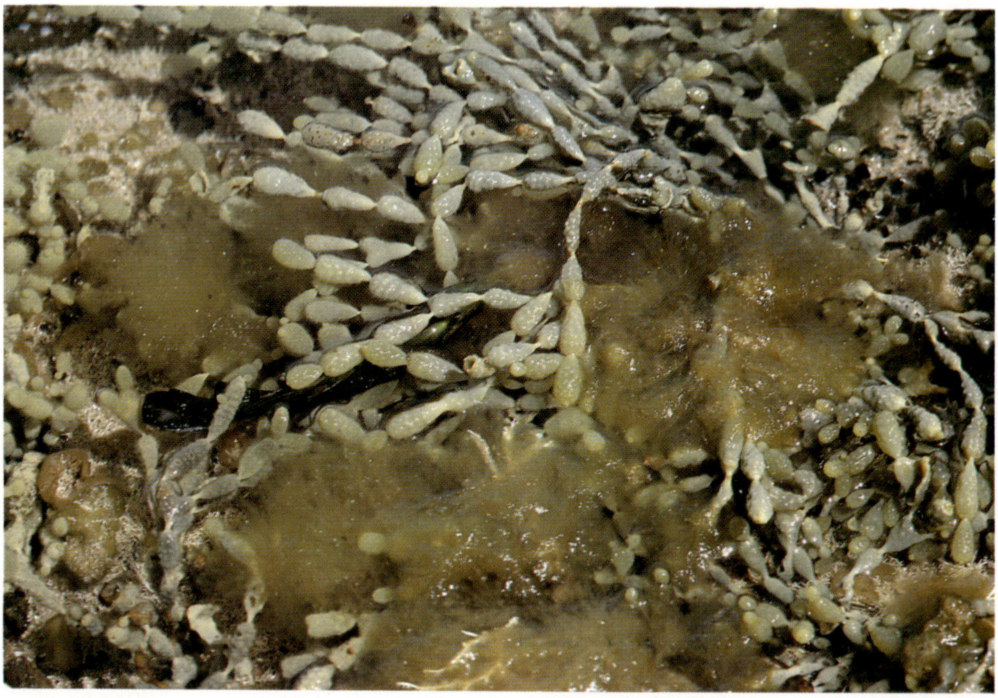

**Top:** Plate 90 (x 4)

*Ectocarpus fasciculatus* (Ectocarpales)
There are many species of brown algae with a similar filamentous appearance in the order Ectocarpales. A microscope is needed to identify the species with certainty.

**Bottom:** Plate 91

*Ectocarpus fasciculatus* growing on *Hormosira banksii*
The soft tufts of this species frequently grow attached to other seaweeds. Old plants of *Scytosiphon lomentaria* (Plate 101) are often densely covered with *Ectocarpus*.

**Top**: Plate 92 (x 4)

*Ectocarpus fasciculatus* (Ectocarpales)
Plants of this species which grow on the surface of rocks differ somewhat in form from those attached to other algae. *E. fasciculatus* occurs throughout the world in temperate coastal regions.

**Bottom**: Plate 93 (x 1)

*Ralfsia* sp. (Chordariales)
Crust-forming brown algae are common on intertidal rocks of southern coasts. Growth occurs at the light-coloured margin, and the crust expands over the rock surface. Some specimens are rather more broken-up in appearance, a result of grazing by herbivorous shore animals such as limpets.

**Top:** Plate 94

*Notheia anomala* (Chordariales)
*N. anomala* grows only attached to *Hormosira banksii*, with which it has a semi-parasitic relationship. The dark, much-branched fronds of *Notheia* are most abundant where *Hormosira* remains wet at low tide—for example, around the edge of tide pools. It can be found at all times of year.

**Bottom:** Plate 95 (x 2)

*Chordaria cladosiphon* (Chordanales)
Superficially this species is rather similar to *Nothheia anomala* but many differences can be seen at the microscopic level, and it does not grow on *Hormosira*. *C. cladosiphon* occurs in relatively sheltered localities in the lower intertidal and upper subtidal zones during spring and summer in south-eastern Australia.

**Top:** Plate 96 (x ⅔)

*Splachnidium rugosum* (Chordariales)
This southern-hemisphere species has a most distinctive form. The thick finger-like projections are filled with a gelatinous substance. *S. rugosum* grows in the intertidal zone of temperate coasts, on rock surfaces which receive moderate to strong wave action. When exposed at low tide, the plants hang down on the rock face.

**Bottom:** Plate 97 (x 1½)

*Petrospongium rugosum* (Chordariales)
Like many species in the Chordariales, *Petrospongium* has a gelatinous texture and is easily squashed. The species characteristically forms a wrinkled, dark brown, spongy crust on the rock surface. It grows in the intertidal zone and is most abundant in winter and spring in south-eastern Australia.

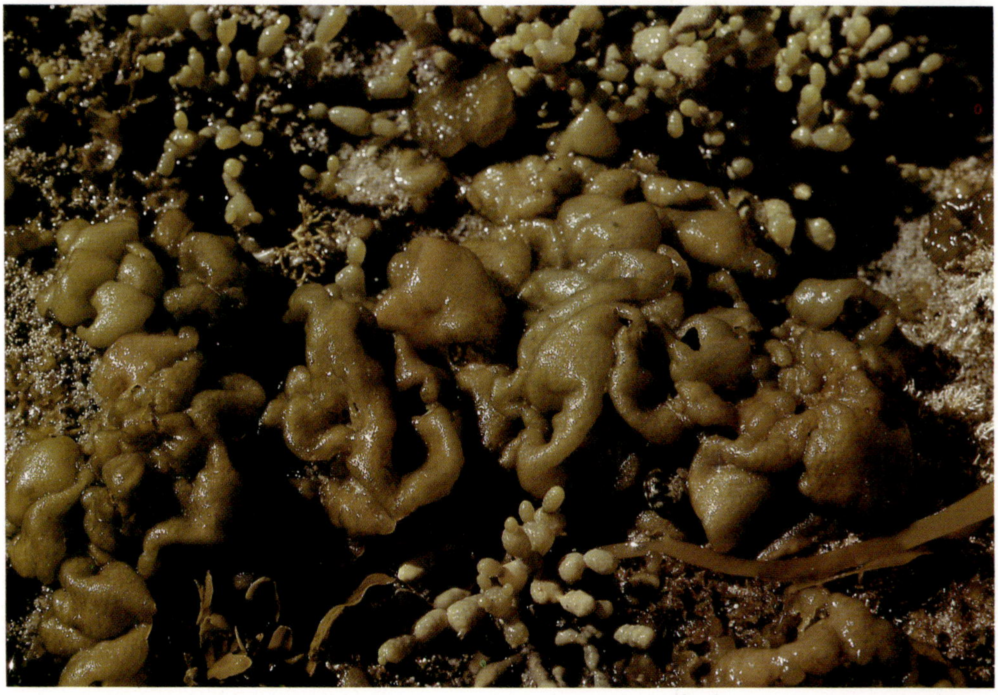

**Top:** Plate 98 (x 1)

*Leathesia difformis* (Chordariales)
This alga is distinguished by a very slimy texture when it is squashed between the fingers. The form is superficially similar to that of *Colpomenia sinuosa. L. difformis* grows in the intertidal zone and can be found for most of the year in southern Australia. In Victoria it is common in late spring and summer. It is a world-wide temperate species.

**Bottom:** Plate 99 (x ½)

*Colpomenia sinuosa* (Scytosiphonales)
There are two species of *Colpomenia* in southern Australia, growing in the intertidal and subtidal zones. They can be found all year round but *C. sinuosa* tends to be more common in summer, and *C. peregrina* is more abundant in winter. They are rather similar in appearance but *C. sinuosa* has a crisper texture. *C sinuosa* occurs on temperate and tropica coasts of Australia whereas *C. peregrin* is a temperate species.

**Top:** Plate 100 (x 1)

*Hydroclathrus clathratus*

(Scytosiphonales)

Small plants of this species resemble a holey *Colpomenia*; large plants can become net-like. *Hydroclathrus* is mainly found in tropical and warm-temperate regions of the world, but it occurs also in southern Australia. It grows in the lower intertidal and upper subtidal zones of the shore.

**Bottom:** Plate 101 (x ⅔)

*Scytosiphon lomentaria*

(Scytosiphonales)

The unbranched tubular fronds of *Scytosiphon lomentaria* occur in a wide range of habitats in temperate coastal waters throughout the world. They may grow intertidally or subtidally, attached to rocks or other seaweeds, and even to limpets. In southern Australia *Scytosiphon* is largest (up to 25 cm long) and most abundant towards the end of winter, but smaller plants may be found throughout the year.

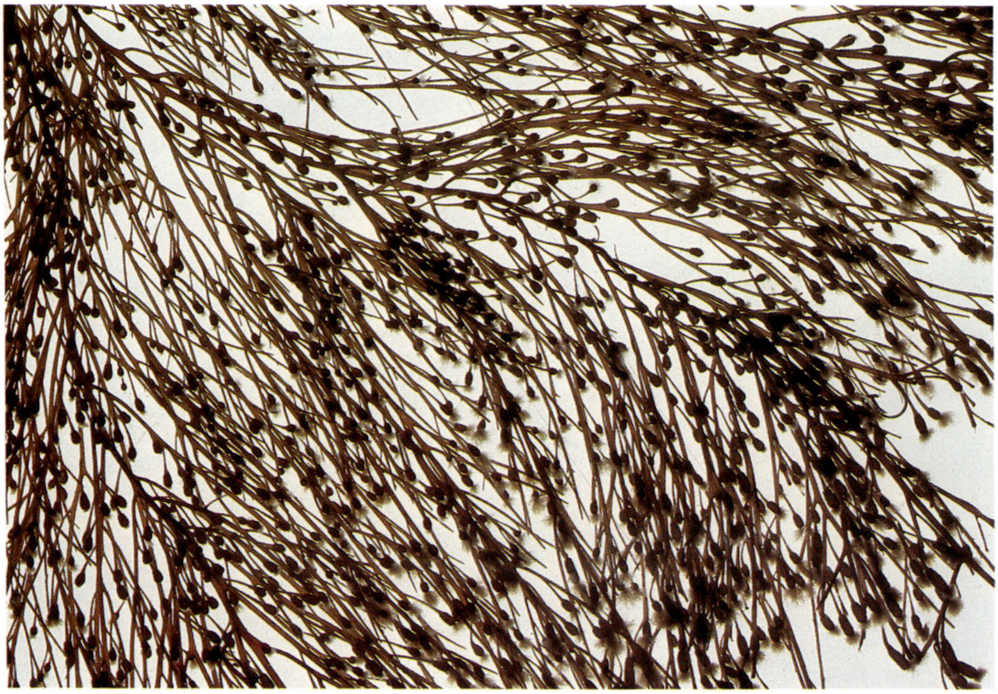

**Top:** Plate 102 (x ¾)

*Bellotia eriophorum* (Sporochnales)
This species is easily recognised by its distinctive brush-like appearance. The conspicuous soft hair tufts are characteristic of the order Sporochnales, a group of about seven genera which are mainly found in the southern hemisphere. *B. eriophorum* is a deep-water alga which may be found cast up as drift along the shores of south-eastern Australia.

**Bottom:** Plate 103 (x 2)

*Sporochnus scoparius* (Sporochnales)
The characteristic hair tufts occurring at the tips of branches are much smaller than they are in *Bellotia eriophorum.* Like *Bellotia*, *S. scoparius* generally grows in the subtidal zone of southern Australian shores. There are several species of *Sporochnus* in Australia.

**Top:** Plate 104 (x ⅓)

*Perithalia caudata* (Sporochnales)
This plant, with dense stiff branches resembling a bunch of twigs, can be up to 60 cm long. It lacks conspicuous hair tufts. *P. caudata* grows subtidally on southern Australian coasts. In this drift specimen, pink encrusting coralline algae can be seen on its main branches and on the piece of rock it was growing on. The red colour of some branches is due to decay.

**Bottom:** Plate 105 (x 2½)

*Halopteris gracilescens* (Sphacelariales)
*H. gracilescens* is a common inhabitant of tide pools on southern Australian shores, and is often found with other algae attached to it. Species of *Halopteris* have slender, rather stiff branches, usually dark brown and becoming pale towards the tips. The dark spot at the very tip of each branch is the large apical cell, characteristic of the order Sphacelariales.

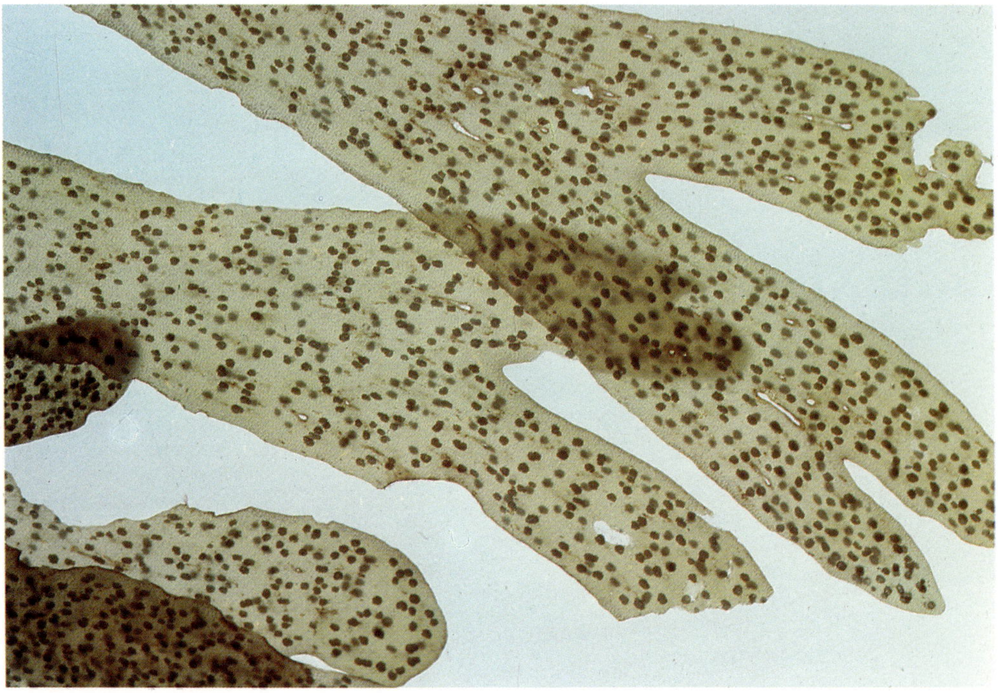

**Top:** Plate 106 (x 3)

*Dictyota dichotoma* (Dictyotales)
This species is found all over the world in a wide range of subtidal habitats. The fronds are often bluish iridescent under water. *D. dichotoma* is very variable in size and form as the plants shown here demonstrate. The small plants are 5 cm tall, while the large plant is 20 cm tall with much wider fronds. The consistent characteristic of the species is the even width of the fronds and the regular branching into equal parts.

**Bottom:** Plate 107 (x 4)

*Dictyota dichotoma*
As with all other members of the Dictyotales, fertile cells are scattered or in patterns over the frond surface. A fertile female plant is illustrated. Several other species of *Dictyota* occur in Australia.

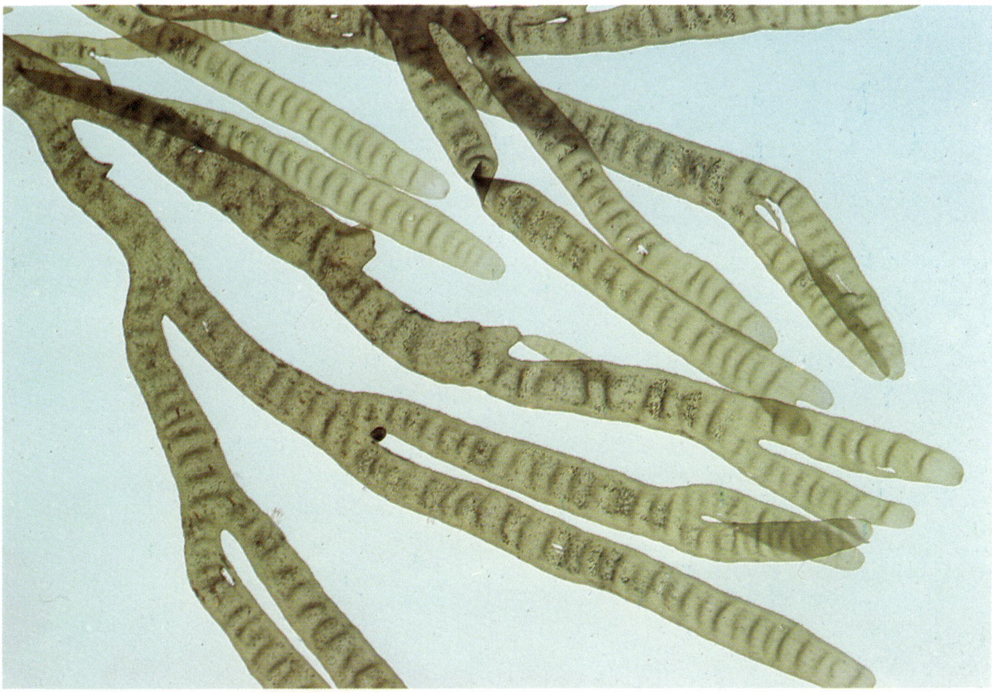

**Top:** Plate 108 (x 2)

*Dilophus marginatus* (Dictyotales)
There are several species of *Dilophus* in Australia, but *D. marginatus* is one of the species which is widespread in both temperate and tropical subtidal areas. *D. marginatus* is distinguished by the banded and wrinkled form of its fertile fronds, which are regularly branched and up to 20 cm in length.

**Bottom:** Plate 109 (x 1½)

*Lobospira bicuspidata* (Dictyotales)
This temperate species occurs only in Australia, and is found in shaded intertidal pools and subtidal habitats. As the name suggests, the frond is spirally twisted, with tooth-like projections on the edges of the fronds. The plants, which grow to 35 cm in length, characteristically have hooked tendrils near the base.

**Top:** Plate 110 (x 1)

*Pachydictyon paniculatum* (Dictyotales)

Only two species of this genus occur in Australia, the larger and more common being illustrated. *P. paniculatum* occurs in temperate subtidal habitats, growing on rock or other seaweeds. The very narrow fronds; which are up to 30 cm long, are often intertwined with each other, and form a soft mass of branches.

**Bottom:** Plate 111 (x 1)

*Padina fraseri* (Dictyotales)

Species of the genus *Padina* are found world-wide, most commonly in tropical and subtropical areas. The southern Australian species, *P. fraseri*, is common in the lower intertidal zone and in rock pools. The fan-shaped fronds are up to 12 cm in height and are lightly calcified.

**Top:** Plate 112 (x ⅓)

*Dictyopteris muelleri* (Dictyotales)
*Dictyopteris* is another widespread genus in Australia. The species illustrated is however restricted to southern Australia in subtidal rocky shore habitats.

**Bottom:** Plate 113 (x 2)

*Dictyopteris muelleri*
In common with all species of the genus, *D. muelleri* has a midrib, but it is distinguished by the broad fronds and the spotted appearance, due to microscopic tufts of hairs. The fronds arise from a narrow base and are up to 36 cm long.

**Top:** Plate 114 (x ½)

*Zonaria angustata* (Dictyotales)
Several species of the genus *Zonaria* are found in intertidal pools and subtidal habitats of southern Australian shores. *Z. angustata* is one of two common species which are not easily distinguished. The branched fronds are narrow (2–5 mm wide) and up to 15 cm in height.

**Bottom:** Plate 115 (x 2½)

*Zonaria spiralis* (Dictyotales)
This species occurs in subtidal habitats in southern Australia. Characteristically it has a tufted habit with narrow (about 2 mm wide), regularly twisted fronds which are up to 15 cm high. The species is the least common of the Zonarias illustrated in this book.

**Top:** Plate 116 (x 1)

*Zonaria* sp. (Dictyotales)
Juvenile plants of all the Zonarias look very similar, and they are common in intertidal localities. The plants form groups of overlapping small broad fronds about 2–4 cm in height. Another brown, *Lobophora variegata*, has similar overlapping fronds and also occurs throughout Australia. It can only be distinguished by microscopic examination.

**Bottom:** Plate 117 (x 1)

*Chlanidophora microphylla* (Dictyotales)
This species occurs in subtidal habitats on rocky shores of southern Australia. As with many of the Dictyotales group, this species is found only in Australia. The species has straight, narrow (to 1·5 mm) branches, and the fronds are up to 16 mm long.

**Top:** Plate 118

*Ecklonia radiata*
Around southern Australian coasts *Ecklonia radiata* is a common and sometimes dominant species just below the level of low tide. Smaller plants also occur in tide pools.

**Bottom:** Plate 119 (x ½)

*Ecklonia radiata*
The fronds of *E. radiata* may vary quite considerably in appearance, according to their degree of dissection. The corrugated blades often have spines on the surface as well as along the margins.

**Top:** Plate 120

*Macrocystis angustifolia* (Laminariales) This large kelp grows subtidally in south-eastern Australia and may form dense communities, with plants growing up to 10 m in length. The long slender stem-like structures (stipes) are buoyed up to the surface by air bladders at the base of the leafy fronds. In some countries *Macrocystis* is harvested commercially for the extraction of alginate, a stabilising agent used in many manufactured products.

**Bottom:** Plate 121 (x ½)

*Macrocystis angustifolia*
The massive 'holdfast', which anchors the plant to the sea floor, is much branched and intertwined, and gives rise to numerous stipes. Tangled drift specimens like the one pictured are often found on the beach.

**Top:** Plate 122

*Durvillaea potatorum* (Fucales)
At a good low tide, on rocky shores of southern Australia that are exposed to heavy wave action and surf, the brown leathery fronds of the 'bull kelp' *D. potatorum* can be seen. On King Island in Bass Strait drift plants are collected and processed for export to be used in alginate manufacture.

**Bottom:** Plate 123

*Durvillaea potatorum*
The plants grow just above and below low tide level to depths of about 10 m. The massive fan-shaped fronds grow to about 8 m in length, and are more or less divided into narrow segments. The strong flexible stem-like 'stipes' stand partly erect as the fronds are washed about by the sea.

**Top:** Plate 124 (x 1/5)

*Durvillaea potatorum*
This photograph of young plants exposed to the air at extreme low tide shows the smooth fist-shaped 'holdfasts' which attach the plants firmly to the rock.

**Bottom:** Plate 125 (x 1/10)

*Cystophora* is a common genus of many species along southern Australian coasts. Most species grow in the upper subtidal zone and in deep pools, where they may be the most conspicuous seaweeds present. In this photograph, taken at extreme low tide, plants of several species of *Cystophora* form a fringe at the reef edge, being attached to the rock by discoid or conical holdfasts.

**Top:** Plate 126

*Cystophora subfarcinata* (Fucales)
The fronds of *C. subfarcinata*, seen here forming a dense growth in a tide pool, may be up to 2 m long. All species of *Cystophora* have a flattened axis (stem-like part) and rather complex branching. Detailed examination may be necessary to distinguish some of the species.

**Bottom:** Plate 127 (x ⅔)

*Cystophora subfarcinata*
The bushy side branches come from the faces of the flat wavy axis. Air bladders, abundant in this tide-pool plant, are absent where wave action is strong.

**Top**: Plate 128 (x ½)

*Cystophora moniliformis* (Fucales)
The branches arise from the *edges* of the flat, broad (about 1 cm) axis. Air bladders are never present in this species.

**Bottom**: Plate 129 (x ⅔)

*Cystophora platylobium* (Fucales)
The name of this species refers to its distinctive flat lobes. *C. platylobium* is generally found in deeper water than the other species illustrated.

**Top:** Plate 130 (x ⅕)

*Cystophora polycystidea* (Fucales)
Clusters of tapered air bladders occur near the base of the side branches. The bushy fronds grow to about 1 m in length.

**Bottom:** Plate 131 (x 1)

*Cystophora retorta* (Fucales)
The forking of the branchlets is distinctly rounded in this species. Air bladders occur only sparsely.

**Top**: Plate 132 (x ¼)

*Cystophora torulosa* (Fucales)
The flattened axis shows a zig-zag form; the club-shaped fertile branchlets clustered along it give this plant a distinctive appearance. *C. torulosa* is sometimes exposed at low tide.

**Bottom**: Plate 133 (x ⅔)

*Caulocystis uvifera* (Fucales)
Throughout the southern Australian region this is very common in tide pools and the upper subtidal zone. It can be distinguished from *Cystophora* by its cylindrical axis (stem-like part) and branching on all sides, with air bladders developed on the axis, not on the side branches.

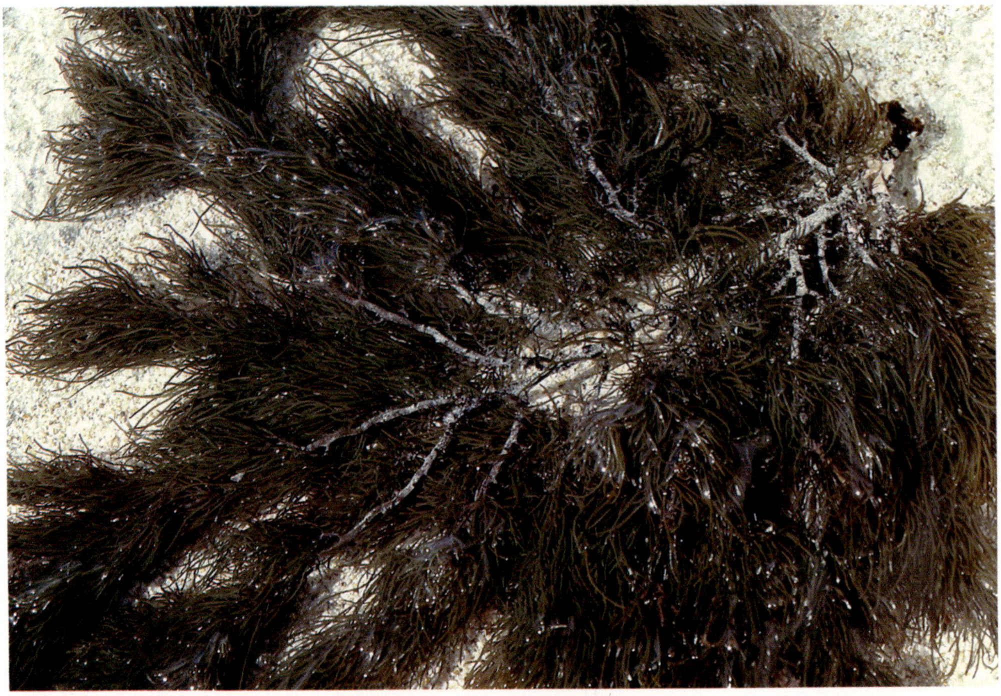

**Top:** Plate 134 (x ½)

*Acrocarpia paniculata* (Fucales)
This genus also is closely related to *Cystophora*. Like *Caulocystis*, it has a cylindrical axis, but it lacks air bladders.

**Bottom:** Plate 135 (x 2)

*Cystoseira trinodis* (Fucales)
This is a widespread species in Australia and the Indo-Pacific region. The plants vary considerably in size and form, with tall thin plants up to 1·5 m high in very sheltered and estuarine waters, or more compact thicker-stemmed plants up to 30 cm high in oceanic reef pools. However, all plants are characterised by smal' peg-like projections on the lower parts o the main branches.

**Top:** Plate 136 (x ⅓)

*Sargassum bracteolosum* (Fucales)
The name of this widespread genus comes from a Spanish word meaning 'floating seaweed'. Both floating and attached species of *Sargassum* are buoyed up in the water by small round air bladders. This large brown alga resembles a flowering plant in its complex branching and distinctive mid-ribbed 'leaves'.

**Bottom:** Plate 137 (x ½)

*Sargassum* sp. (Fucales)
There are many species of *Sargassum* in tropical and temperate Australian waters, growing subtidally and in rock pools. Growth of the upper part of the plant is seasonal and often only the basal leaves, on a short stem, are found. A detailed examination of complete plants is necessary to distinguish the species.

**Top:** Plate 138

*Hormosira banksii* (Fucales)
This is the characteristic species of temperate intertidal habitats on rocky shores and sheltered bays. The fronds may reach 40 cm in length, though the degree of branching, size of the segments, and the total size varies widely. The common name is Neptune's Necklace.

**Bottom:** Plate 139

*Hormosira banksii*
Hot summer days can damage and kill the fronds of seaweeds when they are exposed at low tide. Intense irradiation has turned these *Hormosira* plants a deep reddish-brown colour. This is a common sight on the shore towards the end of summer.

**Top:** Plate 140 (x ⅙)

*Phyllospora comosa* (Fucales)
*Phyllospora* is a large brown seaweed sometimes mistaken for *Macrocystis*. It differs from *Macrocystis* in several ways: the fronds are smooth, narrower and less leaf-like, and the air bladders grow at the end of short stalks, not at the base of the leafy fronds.

**Bottom:** Plate 141 (x ¼)

*Xiphophora chondrophylla* (Fucales)
This fleshy brown seaweed is recognised by its forked branching pattern. The narrow, strap-like fronds repeatedly branch into paired equal sections. It grows in tide pools and in the upper subtidal zone in southern Australia.

**Top:** Plate 142 (x ¾)

*Seirococcus axillaris* (Fucales)

In this distinctive species the reproductive structures appear like many minute buds clustered in the forks of branching fronds. Like most of the other larger brown seaweeds in southern Australia, it grows in the subtidal zone. Plants found cast ashore may be over 1 m in length.

**Bottom:** Plate 143 (x ¾)

*Carpoglossum confluens* (Fucales)

*C. confluens* grows in the subtidal zone; it is found occasionally washed up in the drift. Plants may grow to 1 m or more in length and have a distinctive form.

**Top:** Plate 144 (x ¾)

*Scaberia agardhii* (Fucales)

This seaweed has a unique rough, warty appearance that gives it the name *Scaberia*. It grows in the subtidal zone in southern Australia.

**Bottom:** Plate 145 (x ⅔)

*Turbinaria ornata* (Fucales)

This is a characteristic seaweed of tropical coral reefs, growing in stiff spiny clumps at or just below low tide mark. The plants can occur in several forms: in small clumps of distinctive crown-shaped branches a few centimetres high, or as closely branched fronds up to 25 cm in length.

# GREEN ALGAE
## *(Chlorophyta)*

The most characteristic colour for seaweeds in this group is grass-green, though the shades encountered range from yellow-green to dark or black-green. Unlike the divisions of red and brown algae, which are almost exclusively marine in their occurrence, green algae are found in a wide range of freshwater environments where the majority are microscopic or relatively small 'pond weeds'.
There are fewer species of green seaweeds and they are less spectacular in size and form than the brown and reds. However, some are very common and conspicuous on seashores. Many are tolerant of extreme environmental conditions, including pollution. Thus, seaweeds of this group can be found in the upper intertidal zone of rocky shores, in the vicinity of harbours, in estuaries and at places of freshwater run-off into the sea. Some are also rapid colonisers and grow on new or freshly cleared surfaces in the sea such as boat hulls, piers or solid debris. Other types of green seaweeds are shade-loving and are only found in crevices of rock pools or in the subtidal zone.

**Below:** Plate 146

In winter a luxuriant growth of the green alga *Ulva* may develop on intertidal rock platforms in temperate and subtropical Australia. This reef at Sorrento (Victoria) is dominated by *Hormosira banksii*, with a small form of *Ulva* sp. at higher levels. The white patches indicate mature *Ulva* plants which have released spores.

**Top:** Plate 147

*Ulva* sp. (Ulvales)

Species of *Ulva*, commonly known as Sea Lettuce, have a cosmopolitan distribution and occur in a wide range of subtidal and intertidal habitats, on both rocky shores and sheltered bays. The size and form of the plants vary with changes in environmental factors. *Ulva* grows luxuriantly in waters enriched by organic wastes.

**Bottom:** Plate 148 (x 1)

*Enteromorpha intestinalis* (Ulvales)

This cosmopolitan alga is very common in Australia. Its unbranched tubes, often irregularly constricted, can vary in length from less than 1 cm to more than 25 cm, and in diameter from thread-like to 2 cm.

**Top:** Plate 149

*Enteromorpha* often forms dense turf-like areas in the intertidal zone. Species of *Enteromorpha* are found in such a wide range of habitats on both sheltered and exposed coasts that they could truly be called weeds of the sea!

**Bottom:** Plate 150

*Enteromorpha linza* (Ulvales)
This is another cosmopolitan species. It looks similar to an *Ulva* but can be easily distinguished where it tapers towards the base: this part of the plant is hollow. The pale areas of the plants in this photograph have resulted from the release of reproductive cells into the surrounding seawater.

**Left:** Plate 151 (x 1)

*Chaetomorpha darwinii* (Cladophorales)
This beautiful alga is common in rock pools of temperate Australia. The unbranched chains of large cells, up to 8 mm in diameter, grow on rock or are very often attached to other seaweeds.

**Above:** Plate 152 (x 2)
*Chaetomorpha darwinii* (Cladophorales)

**Top:** Plate 153 (x 1)

*Cladophora rugulosa* (Cladophorales)
Numerous tufted plants of this recent immigrant are seen in a pool on the lower intertidal zone. *C. rugulosa* was known only from South Africa until it was found thirty years ago at Rottnest Island, Western Australia. Today it is widespread in southern Australia, New South Wales and Queensland. *Cystophora torulosa* appears at the top left of this picture.

**Bottom:** Plate 154 (x 3)

*Cladophora rugulosa* (Cladophorales)
The densely-branched filaments have a coarse and almost brittle texture. The plant is a dark green colour, whereas other *Cladophora* species are usually lighter in colour, and have a softer texture.

**Top:** Plate 155 (x 2)

*Cladophora feredayi* (Cladophorales) The bright green tufts of this species are soft and delicate. There are many species of *Cladophora* in Australia, distinguished, but not easily, on microscopic characters.

**Bottom:** Plate 156 (x 1)

*Apjohnia laetivirens* (Cladophorales) This plant has a firm texture and tree-like appearance, growing to about 15 cm high. It is found in rock pools on rough coasts of temperate Australia. The basal part of each branch, particularly the stout lower 'stalk' (not shown), has a series of rings around it. Encrusting coralline algae and other epiphytes are commonly found on *Apjohnia*.

**Top:** Plate 157 (x 1½)

*Bryopsis* sp. (Derbesiales)
The feathery appearance of this plant and its very soft delicate texture are typical of the genus *Bryopsis*. There are several species found in temperate and tropical Australia, growing on rock or on other plants.

**Bottom:** Plate 158 (x 1½)

*Bryopsis plumosa* (Derbesiales)
This almost cosmopolitan species is found on temperate Australian coasts, inhabiting sheltered localities. It can occur in large masses in the spring months, disappearing during the warmer months of the year.

**Top:** Plate 159 (x 1)

*Caulerpa flexilis* (Caulerpales)
There are many species of *Caulerpa* on Australian coasts. The genus is readily recognised by the creeping 'stems', attached by fine branching filaments to rock, sand or mud, and bearing erect green fronds, variously shaped and branched. The species are distinguished on the form of the erect fronds; all are found subtidally or in pools and channels of the lower intertidal zone.

**Bottom:** Plate 160 (x 1)

*Caulerpa annulata* (Caulerpales)
Specimens of this *Caulerpa* can occasionally be found in drift on south-eastern coasts: this is presumably a deep-water species.

**Top:** Plate 161 (x ⅓)

*Caulerpa cactoides* (Caulerpales)
This distinctive, bulky species, named as being like a cactus, is common in rock pools and channels on open coasts along southern Australia. Unlike most other species of *Caulerpa*, it often has encrusting coralline algae on it.

**Bottom:** Plate 162 (x 2)

*Caulerpa cactoides*

**Top:** Plate 163 (x ⅓)

*Caulerpa brownii* (Caulerpales)

The erect fronds of this species are generally unbranched. The dense branchlets give a uniform cylindrical appearance to the fronds, which are of variable length and up to 1 cm in diameter. This is a common species in temperate Australian waters.

**Bottom:** Plate 164 (x 1)

*Caulerpa geminata* (Caulerpales)

In rock pools of the lower intertidal zone on open coasts this plant is only 1–2 cm high, but in calm situations or in deep water it grows higher. The size and arrangement of the small oval branchlets are also variable. *C. geminata* is found in temperate and tropical regions.

**Top:** Plate 165

*Caulerpa longifolia* (Caulerpales)
As its name indicates, the unbranched fronds of this temperate species are very long. In these plants, growing on the edge of the rock platform of an open coast, the branchlets arise in 4 to 6 rows along the fronds.

**Bottom:** Plate 166 (x ¾)

*Caulerpa longifolia*
The form of the fronds is rather variable: in these plants from calm water the branchlets are curved, and not in distinct rows.

**Top:** Plate 167 (x ⅔)

*Caulerpa obscura* (Caulerpales)
The dark green frond has irregularly arranged side branches, covered with long thin branchlets, and has a more bushy form than the other species illustrated. *C. obscura* grows on open rough coasts: this specimen was collected as drift.

**Bottom:** Plate 168 (x 1)

*Caulerpa remotifolia* (Caulerpales)
The flattened fronds have pointed, widely separated branchlets. *C. remotifolia* is found in calm sheltered bays in southern Australia.

**Left:** Plate 169 (x 3)

*Caulerpa racemosa* var. *clavifera* (Caulerpales)

*C. racemosa* occurs throughout the tropics and subtropics where it is probably the most common species of the genus. The erect branches in var. *clavifera* are up to 1·5 cm tall and resemble miniature branches of grapes. Each branchlet expands rapidly but smoothly from the slender stalk to the globular tip.

**Left:** Plate 170 (x 1⅓)

*Caulerpa vesiculifera* (Caulerpales)

There are several species of *Caulerpa* which are rather similar to *C. vesiculifera* found in southern Australia.

**Top:** Plate 171 (x 1)

*Halimeda* sp. (Caulerpales)
*Halimeda* is one of the most characteristic and common algae of tropical and subtropical regions throughout the world, and contributes substantially to the sediments of coral reefs. Plants are up to 20 cm or more in length, and may grow on rock or 'take root' in sediment. The genus is characterised by its calcified disc- or barrel-shaped segments, which are linked together by uncalcified nodes.

**Bottom:** Plate 172 (x ½)

*Chlorodesmis fastigiata* (Caulerpales)
Commonly called Turtle Weed, this is a conspicuous tropical plant in the Indo-West Pacific region, usually on coral reef rock, in both the intertidal and subtidal zones. The brilliant green balls of repeatedly forked filaments are up to 10 cm high.

**Top:** Plate 173 (x 1½)

*Codium duthiae* (Codiales)
The microscopic structure of the genus *Codium* gives a spongy texture and appearance of velvet pile to the plant surface. There are many species in Australia, of varied external form. In *C. duthiae* the cylindrical forked branches are often slightly flattened, and up to 60 cm long.

**Bottom:** Plate 174 (x ⅓)

*Codium fragile* (Codiales)
The cylindrical, forked, dark green fronds of *C. fragile* grow to 30 cm long. When the plant is under water, fine hairs can be seen over the surface of the branches. This is a species of temperate regions, found subtidally and in intertidal pools, often on rough coasts.

**Top:** Plate 175 (x ⅙)

*Codium fragile* (Codiales)
Small red algae are often found growing on *C. fragile*, giving a pink colour to the fronds.

**Bottom:** Plate 176 (x ⅓)

*Codium fragile* (Codiales)
Particles of sand have settled in the spongy surface of this plant exposed at low tide. There are several species of *Codium* with similar forked branching, distinguished mainly on microscopic characters.

**Top:** Plate 177 (x ¾)

*Codium lucasii* (Codiales)
The thin, very dark green spreading lobes have a firm texture and close 'pile', and are attached tightly to rock surfaces in pools and the lower intertidal zone. This species occurs in temperate and sub-tropical Australia.

**Bottom:** Plate 178 (x ¾)

*Codium pomoides* (Codiales)
The name of this species means 'resembling an apple'. The dark, very firm plants are up to 10 cm in diameter; they are found on shaded rock surfaces in pools and crevices, and subtidally, on rough coasts in southern Australia.

**Top:** Plate 179 (x ¾)

*Codium mamillosum* (Codiales)
This plant resembles *C. pomoides* but it has a softer texture and coarser 'pile' and is light green in colour. It is found mainly in calm localities on southern coasts.

**Bottom:** Plate 180 (x ¾)

*Codium spongiosum* (Cordiales)
The spreading, undulating lobes have a soft spongy texture. This species is temperate and tropical in distribution, and found in relatively calm conditions.

**Top**: Plate 181 (x 1)

*Dictyosphaeria sericea*

(Siphonocladales)

These cup-shaped plants are very firmly attached to rocks by their wide basal parts. They are found in pools and channels of the lower intertidal zone, often on vertical rock faces. This species occurs on open coasts of southern Australia; other species of *Dictyospheria* are tropical.

**Bottom**: Plate 182 (x 3)

*Valonia ventricosa* (Siphonocladales)

This tropical plant consists of a large, dark green, rounded cell up to 3 cm in diameter attached by minute basal rhizoids. The surface of the plant usually has an iridescent sheen, which is often obscured by crusts of epiphytic coralline red algae. The plant occurs in rock crevices or among rubble.

**Right**: Plate 183 (x 1¼)

*Bornetella* sp. (Dasycladales)
The genus *Bornetella* occurs predominantly in the tropical Indo-West Pacific region, where it may be abundant on coral rubble in the intertidal zone. Some species are globular, while others (as illustrated) are club-shaped and up to 2·5 cm tall. A slender central axis with whorls of fine lateral branches is present internally. The plants are distinctly green in colour but are lightly calcified.

**Right**: Plate 184 (x 4)

*Polyphysa peniculus* (Dasycladales)
This delicate little alga is found on sheltered sandy or muddy shores about low tide level, where it grows on old shells. The plants are up to 4 cm high. *P. peniculus* is found in temperate Australian waters; most other species of *Polyphysa* and the closely related genus *Acetabularia* are tropical and subtropical.

# SEA-GRASSES
## *(Marine Angiosperms)*

In many bays, estuaries and other shallow, relatively sheltered parts of the sea, the flowering plants known as sea-grasses are an important constituent of the marine vegetation. They form large underwater meadows which are breeding grounds for fish and invertebrates, and also stabilise the sand or mud surfaces. Sea-grasses are not related to the algae: they have true stems and leaves, inconspicuous flowers, and form fruits and set seeds like flowering plants on land. They are, however, quite distinct from the land grasses. Great masses of sea-grasses often pile up on sandy shores after storms.

**Below:** Plate 185 (x 1/3)

*Amphibolus antarctica* Sea Nymph
This plant occurs only in Australia, in the cooler parts of the coast from Victoria to Western Australia and northern Tasmania. It is frequently colonised by algae (see Plate 11) which grow on the long stems and tough persistent leaves. The seedlings germinate on the mother plant, are later distributed and become attached to the seafloor by a 'grappling apparatus' formed from the fruits.

**Below**: Plate 186 (x ½)

*Zostera muelleri* Eelgrass

The genus *Zostera* is very widespread throughout the world. This Australian species occurs along the south-eastern coast and in Tasmania. Its narrow strap-shaped leaves are often seen exposed at low tide on sheltered mud and sand banks.

# FURTHER READING


Clayton, M. N. & King, R. J. (eds) *Marine Botany: an Australasian Perspective*. Longman Cheshire, 1981. (A comprehensive textbook for Australian and New Zealand students.)

Cribb, A. B. & J. W. *Wild Food in Australia*. Collins, 1974. (Includes a chapter on seaweeds as a source of food.)

Dakin, W. J. *Australian Seashores*. Angus and Robertson, revised edn, 1973. (A general account, mostly about animals, but includes a chapter on plants.)

# INDEX TO GENERA AND SPECIES